Words to Know

color

fins

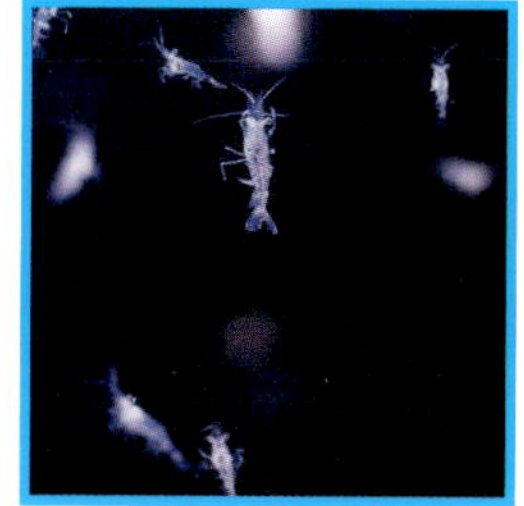

plankton

seahorse

tails

This is a **seahorse**.

UNDER THE SEA ANIMALS
SEAHORSES

Amy Culliford

TABLE OF CONTENTS

A Pelican Book

Teaching Tips for Caregivers and Teachers:

Research shows that one of the best ways for students to learn a new topic is to read about it.

Before Reading

- Read the title and predict what the book will be about.
- Read the "Words to Know" and discuss the meaning of each word.
- Read the back cover to see what the book is about.

During Reading

- When a student gets to a word that is unknown, ask them to look at the rest of the sentence to find clues to help with the meaning of the unknown word.
- Motivate students with praise and encouragement.

After Reading

- Discuss the main idea of the book.
- Ask students to give one detail that they learned in the book.

Sight Words

a
all
any
be
big
can
eat
have
is
little
long
or
some
this

seahorse

A seahorse can be any **color**.

color

Seahorses can be big or little.

All seahorses eat **plankton**.

plankton

Some seahorses have long **tails**.

tail

All seahorses have **fins**.

fin

Index

Written by: Amy Culliford
Design by: Under the Oaks Media
Series Development: James Earley
Editor: Kim Thompson

Photos: Wildstrawberry: cover; Arunee Rodley: p. 5; Vojce: p. 7; Juliet Lisa Rose: p. 9; M-Production: p. 11; Andrea Izzotti: p. 13; Tim_Walters: p. 15

Library of Congress PCN Data
Seahorses / Amy Culliford
Under the Sea Animals
ISBN 978-1-63897-068-2 (hard cover)
ISBN 978-1-63897-154-2 (paperback)
ISBN 978-1-63897-240-2 (EPUB)
ISBN 978-1-63897-326-3 (eBook)
Library of Congress Control Number: 2021945240

Printed in the United States of America.

Seahorse Publishing Company
www.seahorsepub.com

Published in the United States
Seahorse Publishing
PO Box 771325
Coral Springs, FL 33077